Nature's Mysteries

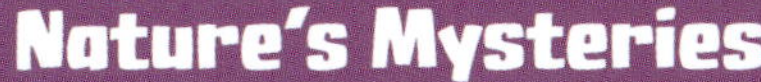

PERU'S RAINBOW MOUNTAIN

by Rachel Hamby

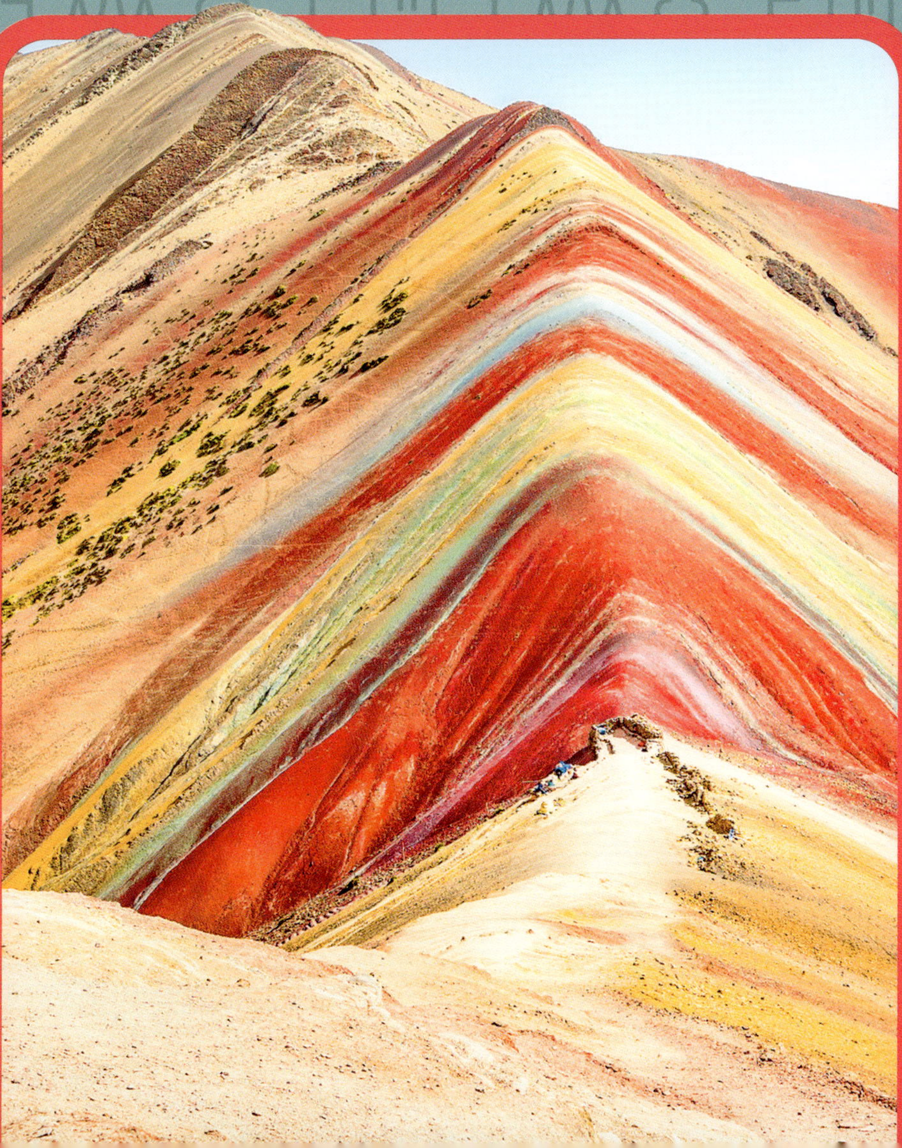

abdobooks.com

Published by Pop!, a division of ABDO, PO Box 398166, Minneapolis, Minnesota 55439.

Printed in the United States of America, North Mankato, Minnesota.

102020
012021

THIS BOOK CONTAINS RECYCLED MATERIALS

Cover Photo: iStockphoto
Interior Photos: iStockphoto, 1, 5, 8, 12, 14–15 (top), 20; Shutterstock Images, 6, 7, 9, 14–15 (bottom), 16–17, 19, 21, 22, 23, 25, 26, 28, 29; Carmen Gabriela Filip/Alamy, 11; Tino Plunert/dpa/picture-alliance/Newscom, 27

Editor: Alyssa Krekelberg
Series Designers: Candice Keimig, Victoria Bates, and Laura Graphenteen

Library of Congress Control Number: 2020940298

Publisher's Cataloging-in-Publication Data

Names: Hamby, Rachel, author.

Title: Peru's Rainbow Mountain / by Rachel Hamby

Description: Minneapolis, Minnesota : POP!, 2021 | Series: Nature's mysteries | Includes online resources and index

Identifiers: ISBN 9781532169212 (lib. bdg.) | ISBN 9781532169571 (ebook)

Subjects: LCSH: Andes--Juvenile literature. | Mountains--Juvenile literature. | Soil mineralogy--Juvenile literature. | Geology--Peru--Juvenile literature. | Curiosities and wonders--Juvenile literature. | Mystery--Juvenile literature.

Classification: DDC 910.02--dc23

WELCOME TO DiscoverRoo!

Pop open this book and you'll find QR codes loaded with information, so you can learn even more!

Scan this code* and others like it while you read, or visit the website below to make this book pop!

popbooksonline.com/rainbow-mountain

*Scanning QR codes requires a web-enabled smart device with a QR code reader app and a camera.

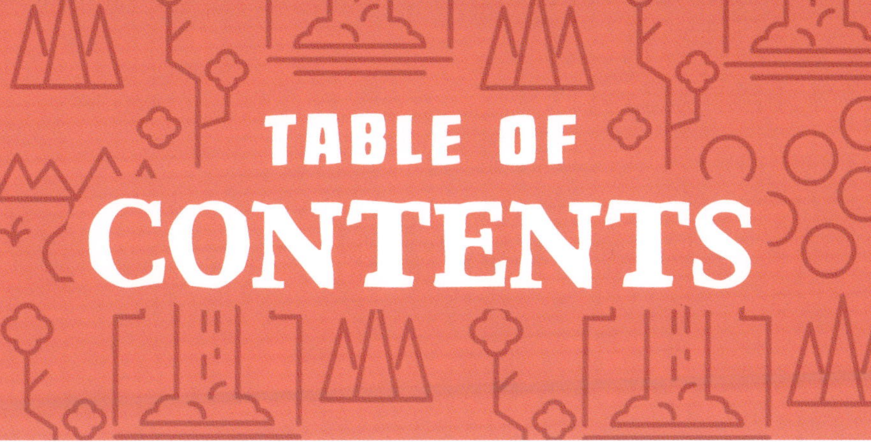

TABLE OF CONTENTS

CHAPTER 1

RAINBOW MOUNTAIN

Hikers in Peru strap on their backpacks. They are on their way to Rainbow Mountain. They walk for miles. Sometimes the trail is steep. The hike is challenging. But the view is amazing.

WATCH A VIDEO HERE!

The trip to Rainbow Mountain can take all day.

The Andes stretches approximately 5,500 miles (8,850 km).

Rainbow Mountain has colorful stripes. Each stripe is a layer of rock. The mountain is part of the Andes. This mountain range is very long. It stretches down the west coast of South America.

WHERE IS RAINBOW MOUNTAIN?

Rainbow Mountain formed millions of years ago. **Glaciers** have been on the Andes for hundreds of years. One glacier covered Rainbow Mountain. But as Earth

Many glaciers in Peru are melting because of climate change.

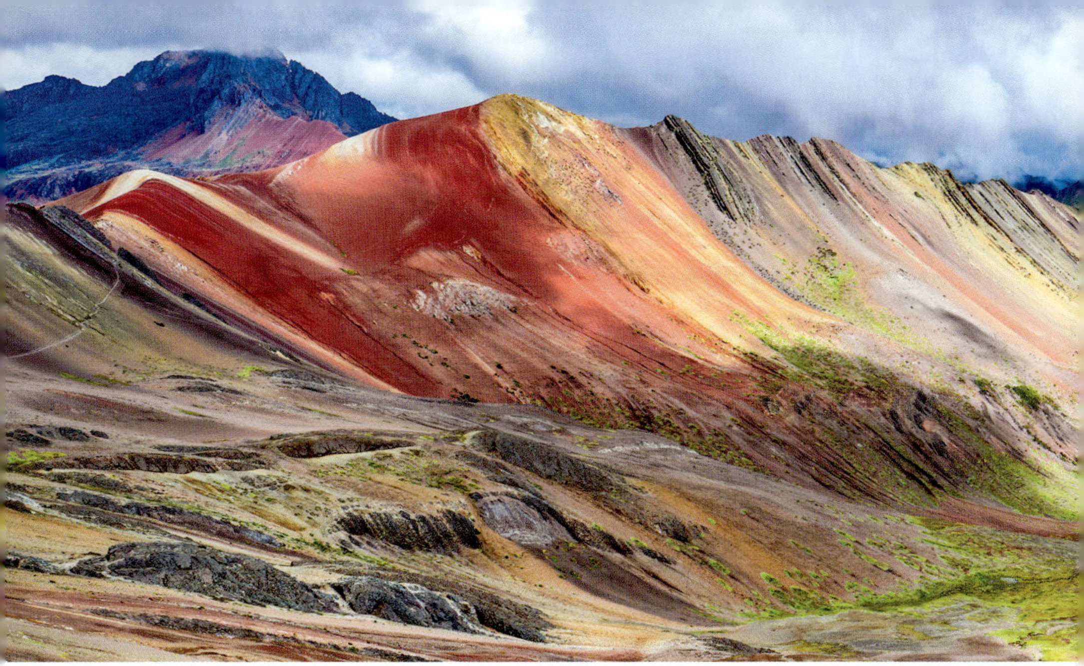

People were amazed when they discovered the colorful mountain.

experienced **climate change**, the glacier melted. The mountain's colors became visible. People first saw them in 2015.

FUN FACT

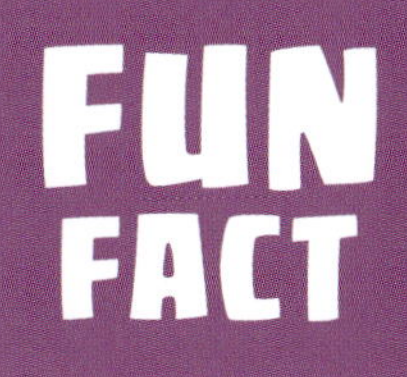

Rainbow Mountain is also known as the seven-colored mountain.

CHAPTER 2

COLORFUL ROCKS

Rainbow Mountain has many colors. Some of these colors are red, yellow, purple, and blue. The stripes are layers of sedimentary rock. Each layer formed at a different time.

LEARN MORE HERE!

People who visit Peru can experience the country's rich culture.

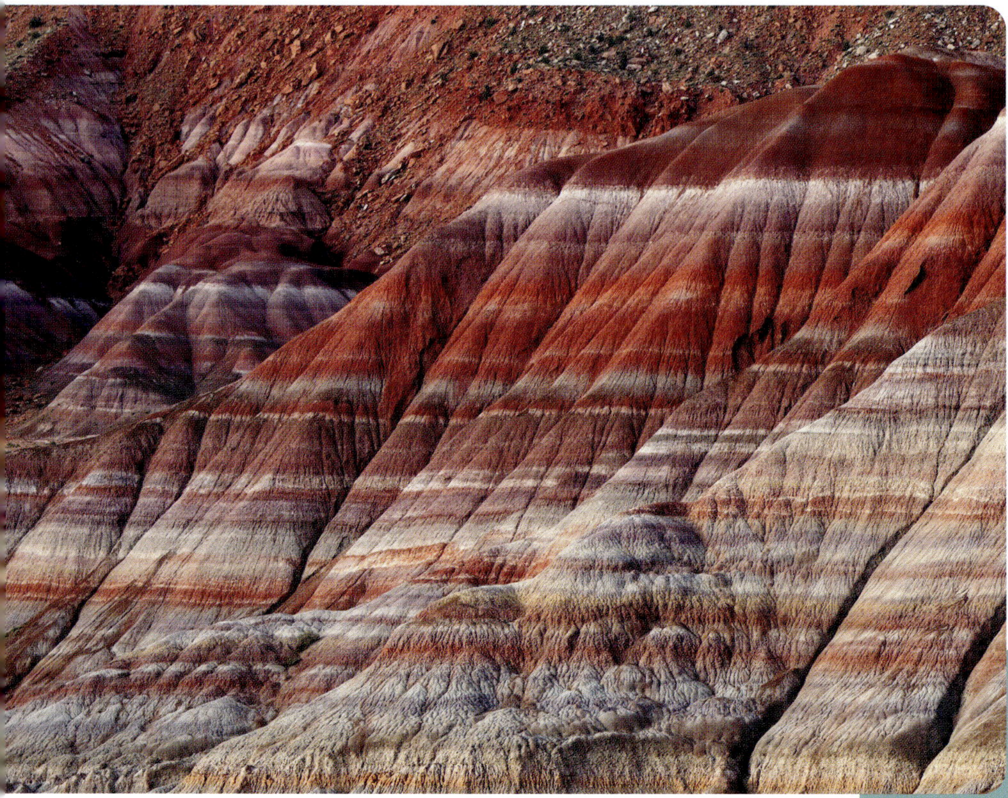

Rock layers around the world tell geologists what areas looked like thousands of years ago.

FUN FACT

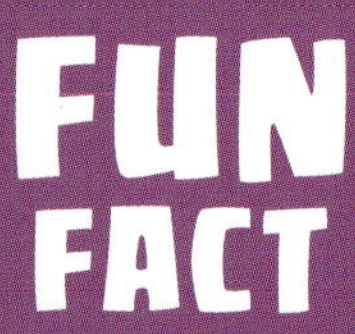

Limestone is found in Rainbow Mountain. This rock forms in warm, shallow water. That means Rainbow Mountain was once underwater!

The rock layers have different **sediments**. Some have clay, sand, or limestone. They help **geologists** learn about the mountain's history. They show what Earth was like at different times.

WHAT ARE SEDIMENTARY ROCKS?

Sedimentary rocks form on Earth's surface. First, large rocks break into small pieces. These pieces are called sediment. They are moved by water and wind. Then, they settle. More sediment rests on top. Rocks at the bottom are smashed together. Water and **minerals** seep in. Everything sticks together. A sedimentary rock is formed.

The mountain's yellow coloring also comes from a type of iron.

Different minerals are found in each layer. They create the mountain's colors. Some layers of minerals are changed by weather. For example, red layers have iron that was exposed to water and oxygen. The iron **oxidized**. The rock turned red.

The top layer of a mineral might be a different color from the layer underneath.

MINERALS IN RAINBOW MOUNTAIN

iron oxide
calcium carbonate

CHAPTER 3

A RISING MOUNTAIN

Long ago, the layers of rock that make up Rainbow Mountain were flat. Tectonic shift pushed them up. This is the movement of Earth's crust. The crust is made up

LEARN MORE HERE!

of tectonic plates. The plates are always moving. Sometimes they slide or crash together. They push up some layers of rock.

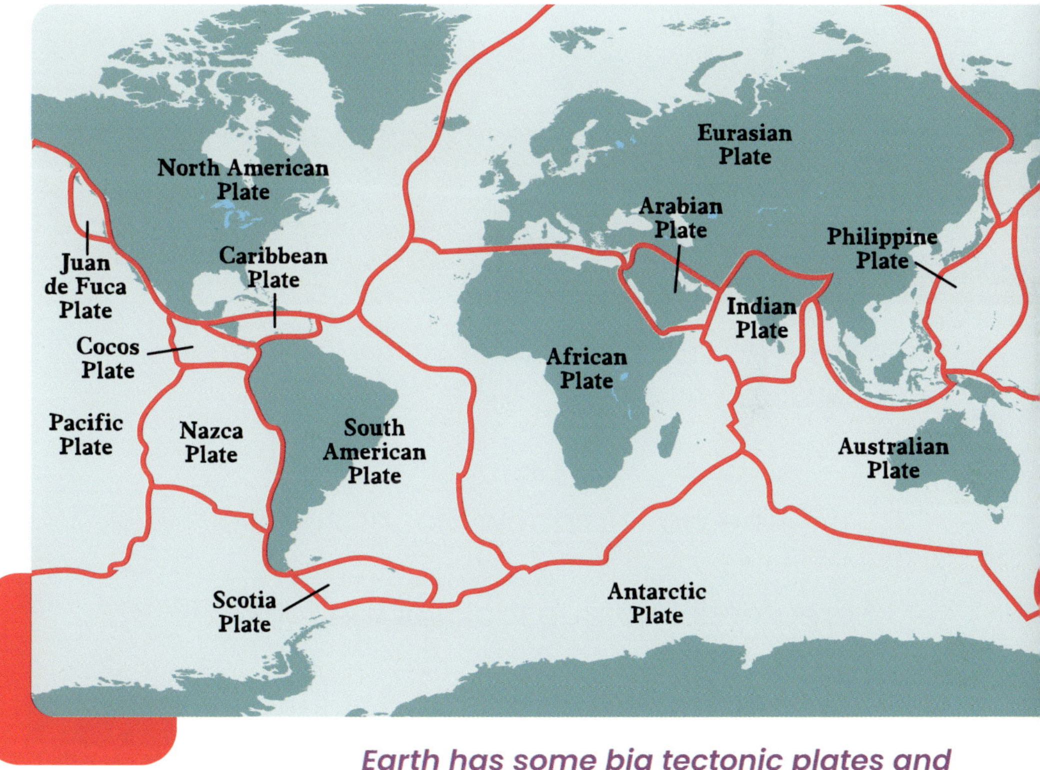

Earth has some big tectonic plates and some small ones.

Tectonic plates crashing against each other formed the Andes.

The way Rainbow Mountain rose created its unique look. As the rock pushed up, the layers tilted sideways. This process took millions of years.

Today, the mountain's **elevation** is 20,945 feet (6,384 m).

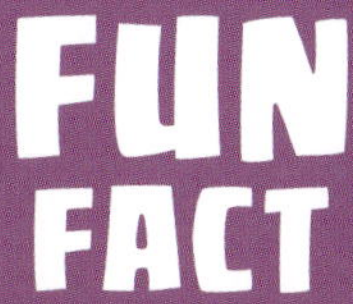

FUN FACT **The movement of tectonic plates can cause earthquakes or make volcanoes erupt.**

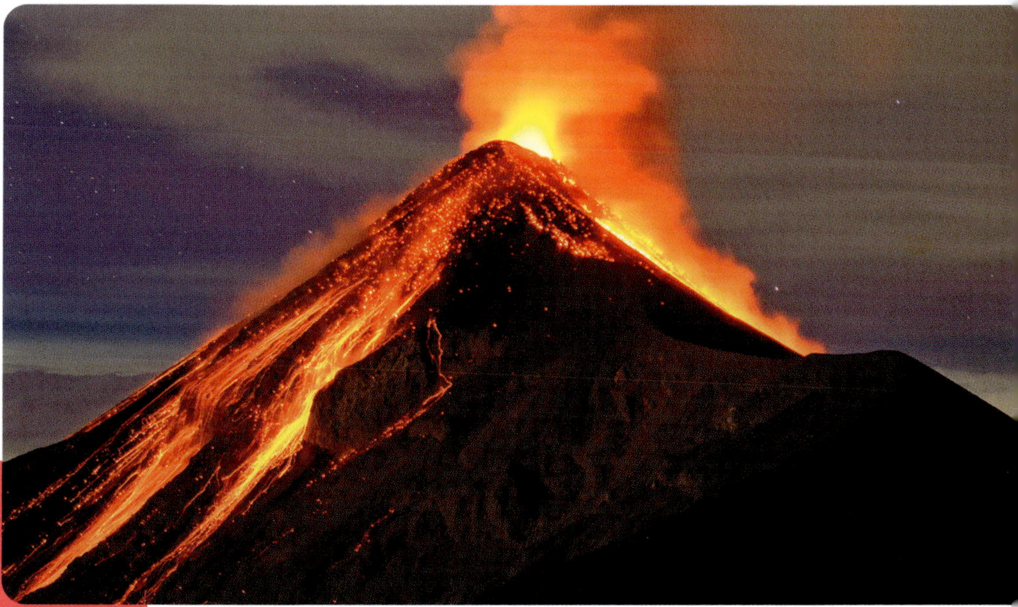

Volcanic eruptions happen when magma is pushed to Earth's surface. When the magma comes out of a volcano, it is then called lava.

Over many years, rainfall can affect mountains.

The mountain was also shaped by weather. Wind and rain broke down rocks on the mountain. **Glaciers** did too. These things wiped away Rainbow Mountain's rocky surface. This is why people can see the mountain's many colors.

Loose rocks can slide down a mountain.

CHAPTER 4

THE FUTURE OF RAINBOW MOUNTAIN

Geologists know that Rainbow Mountain will change over time. Some of these will be natural changes. But people can affect the mountain too. More hikers visit every year. People are learning how this could change the mountain.

COMPLETE AN ACTIVITY HERE!

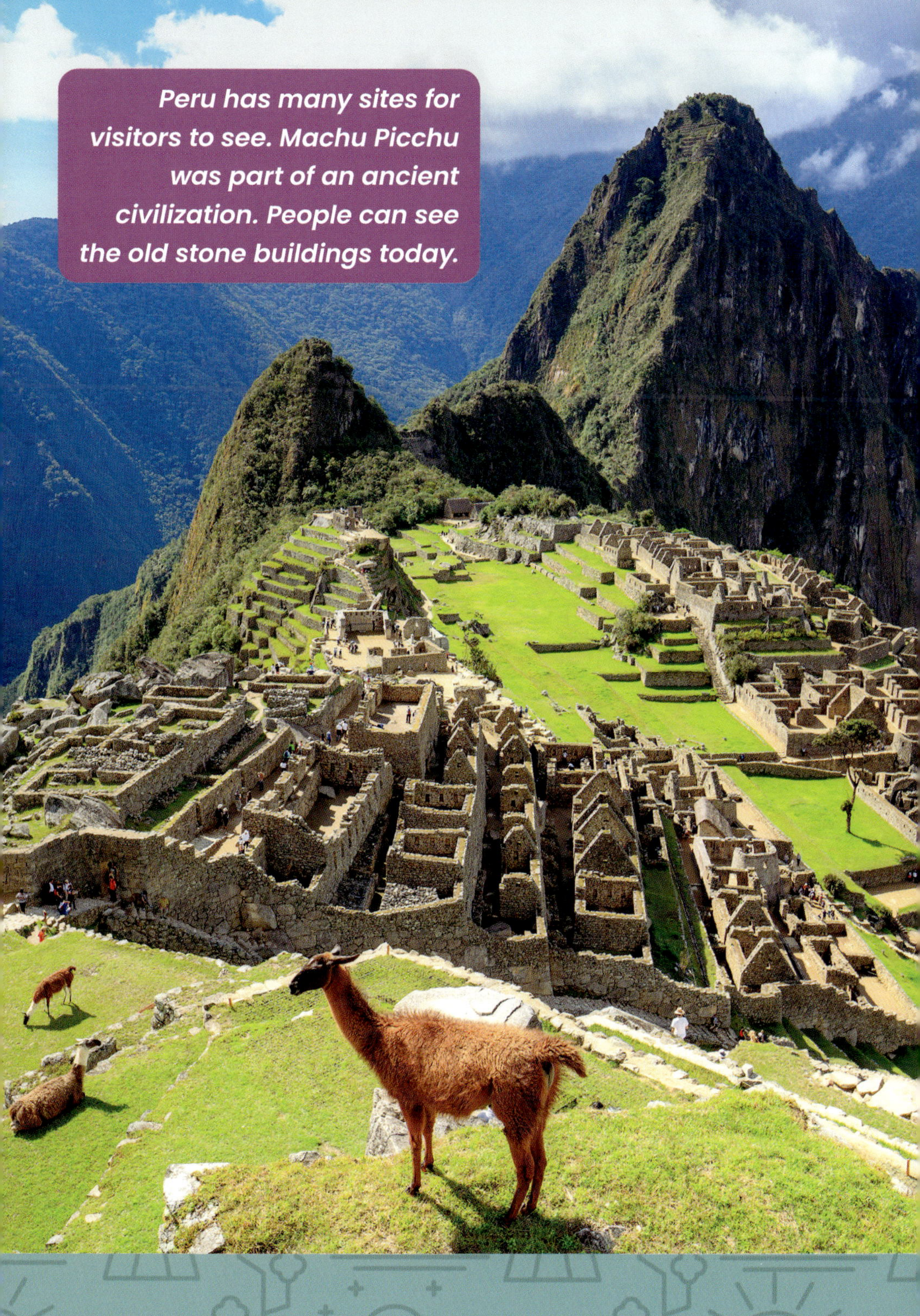

Peru has many sites for visitors to see. Machu Picchu was part of an ancient civilization. People can see the old stone buildings today.

Mining changes how an area of land looks.

Minerals are used for many things. In mining, rock is dug from the earth. The minerals are removed. Mining breaks down the land. Mining companies want the minerals in Rainbow Mountain.

But Peru's government has stopped them.

Miners want to get ahold of Rainbow Mountain's iron and copper.

People in Peru value Rainbow Mountain. It is a new discovery. But people are still learning how to protect it. That way, the colors can be seen for many more years.

The hike to Rainbow Mountain can be hard for some people. But many hikers like the view.

People who visit Rainbow Mountain might also see llamas and alpacas nearby.

FUN FACT

Rainbow Mountain is very popular. Approximately 1,500 people visit it every day.

MAKING CONNECTIONS

TEXT-TO-SELF

Would you want to visit Rainbow Mountain? Why or why not?

TEXT-TO-TEXT

Have you read other books about natural wonders? How is Rainbow Mountain different from other places you've read about?

TEXT-TO-WORLD

By not allowing mining, Peru helps protect Rainbow Mountain. What are some things you can do to help the environment where you live?

GLOSSARY

climate change — a crisis that is causing Earth's weather patterns to change, often including rising temperatures.

elevation — a measurement of how high a place is above sea level.

geologist — a scientist who studies the history of Earth and life through rocks.

glacier — a large body of ice that moves slowly down a slope.

mineral — a nonliving solid that forms naturally.

oxidize — to combine with oxygen.

sediment — bits of rocks, shells, and other things.

INDEX

ONLINE RESOURCES

popbooksonline.com

Scan this code* and others like it while you read, or visit the website below to make this book pop!

popbooksonline.com/rainbow-mountain

*Scanning QR codes requires a web-enabled smart device with a QR code reader app and a camera.